The Smith Sexagesimal System

Using Base-60 to Increase Arithmetic Intelligence

Jean-Michel Smith

RED ANEMONE BOOKS

Copyright © 2005, 2011 by Jean-Michel Smith

All rights reserved.

Published in the United States by Red Anemone Books, Chicago.

ISBN: 978-0-9831888-1-0
eBook ISBN: 978-0-9831888-2-7

First published in the United States: December 2011
First eBook Edition: December 2011

**RED ANEMONE
BOOKS**

redanemone.com

For my father Jean Michael Hurlbert-Smith, R.I.P.
and for my lovely wife Christine Todd,
without whose brilliance, patience, and encouragement
this book would never have come into being.

BOOKS BY JEAN-MICHEL SMITH

FICTION
Autonomy, a novel

MATHEMATICS
S^3: The Smith Sexagesimal System

CONTENTS

	Introduction	vii	
1	Positional (Place-Value) and Integer Base Notation	3	ω
2	Written Sexagesimal Numerals	6	∪
3	Sexagesimal Nomenclature	9	ω
4	Converting Between Sexagesimal and Decimal	21	ᘯ
5	Maxwell Planck's "Natural Units" and Metric-60 Defined	27	ᘲ
6	Planck Temperature and Degrees Prime	29	೪
7	Planck Time and "Tocks"	32	ᔑ
8	Planck Length and "Tocks"	34	ᔓ
9	Planck Mass and "Specks"	36	ᔕ
10	Planck Charge and "Sparks"	37	ᔖ
11	Derived Units	38	ᔗ
12	Fictional Metric-60 Units from the Novel *Autonomy*	41	ᔘ
	Notes	47	ᔙ
	A preview of the novel Autonomy	49	೪

INTRODUCTION

When I wrote my science fiction novel *Autonomy*, I pondered how hyper-intelligent software entities would cope if forced to download into human bodies. How would such beings make the best use of their now-limited thinking capacity? I tossed around a number of ideas on how they might increase human arithmetic intelligence. Ironically, this led me to the ancient base-60 numerical system first used by the Sumerians. A few modern tweaks leveraging contemporary mathematical concepts and notation gave rise to the Smith Sexagesimal System and, ultimately, the related Natural Metric-60 Units of Measure.

Metric-60 is built upon a system of "natural units" first proposed by German physicist Max Planck. Natural units derive from the laws of physics as we currently understand them, rather than arbitrary human definitions. Using natural units leads to some interesting results:

1. The same units are used to measure time and space
2. The same units are used to measure matter and energy

Concepts in physics which are not very intuitive become obvious if Planck units are used to describe them. But the problem with Planck units is their ungainliness. When using decimal numerals, the basic units are too small, or in the case of temperature too large, to be convenient for day-to-day use.

By using the Smith Sexagesimal System and the related Metric-60 system, Planck units don't suffer this problem of scale and become easier to use. Sexagesimal can represent much larger numbers than decimal in few enough digits to be grasped and manipulated by the average human mind. Most of us can only hold three or four digits in our short term memory.[1,2] The fewer digits needed to represent a value, the larger the value that can be mentally manipulated and remembered.

This is best illustrated with a couple of examples. Consider the number 3521. Easy enough to grasp and remember in decimal, but in a lower base such as binary (base-2) it becomes a jumble of digits difficult to commit to memory: 110111000001. Sexagesimal offers a similar advance over decimal, representing the number 3521 with just two digits: ↑ℰ. What's more, the advantage of sexagesimal over decimal grows dramatically with each digit:

1. 777,600,000 in decimal is represented by /00,000 in sexagesimal.

2. 46,655,999,999 in decimal is represented by ዓዓዓ,ዓዓዓ in sexagesimal.

S³: The Smith Sexagesimal System

Another advantage of sexagesimal is even division by 2, 3, 4, 5, 6, 10, 12, 15, 20, and 30, instead of the limited pair 2 and 5 available to decimal. Numerous fractions difficult to represent in decimal can be written elegantly in base-60. For example, the fraction $1/3$ can only be approximated in decimal as $0.33333\cdots$, whereas in sexagesimal $1/3$ is expressed exactly with one digit after the radix point, like so: 0.⚇, meaning (0).(20)$_{60}$. (Sexagesimal symbols are fully explained in the chapter entitled "Written Sexagesimal Numerals".) This advantage is so profound we still use base-60 today, in time (60 seconds in a minute, 60 minutes in an hour), degrees of angle (360 degrees in a circle), and consequently in navigation, astronomy, and numerous other disciplines.

If sexagesimal is so much better than decimal, and the Sumerians were using it in the third millennium BC, why isn't everything in base-60 today? The answer probably has as much to do with history and chance as anything, but the cumbersome cuneiform notation the Sumerians and Babylonians used didn't help. Cuneiform characters were written in clusters of tally marks like so:

𒌋𒌍 [3] = 18

𒌍𒌍 [4] = 36

𒁹𒌋𒌍 [5] = 109

𒌋𒌍𒌋𒌍𒌋𒌍𒌍 [6,7] = 4,274,228

Clearly, cuneiform characters are not easy for the eye to parse, particularly when dealing with more than a few digits.

The ancients also lacked something we take for granted: a standard symbol for "0" (though there is some evidence a dot or a space may have been inconsistently used as an empty place holder).

The advanced sexagesimal system I've developed solves these problems by leveraging the positional notation of our decimal system (including a radix point separating positive and negative powers of 60), unique and simple single-character representations of each digit 0-⅋ (0-59), a nomenclature for naming a vast range of numbers based on powers of 60, and a Metric-60 nomenclature for unit prefixes. The system is designed to be concise, coherent, and above all, easy to understand.

—Jean-Michel Smith
November 2011

PART 1

THE SMITH SEXAGESIMAL SYSTEM

1 – POSITIONAL (PLACE-VALUE) AND INTEGER BASE NOTATION

As discussed in the introduction, the Smith Sexagesimal System is a concise, coherent, and easy to use tool for increasing arithmetic intelligence. How exactly does a base-60 numbering system work? First, it is important to review how our positional base-10 system works. Those already familiar with numerical systems of different bases can skip to the next chapter, "Written Sexagesimal Numerals".

Positional notation is so intuitive that we grasp a fairly complex arithmetic equation with just a glance. We immediately understand 195.02 without having to add 5 to 90 to 100, and finally include the fraction $^2/_{100}$.

Positional notation means that the position of each digit relative to the radix point determines the magnitude of its value. The first place to the left of the radix point represents the digit multiplied by the numerical base taken to the 0th power, the second represents the digit multiplied by the numerical base to the first power, the third represents the digit multiplied by the numerical base squared, and so on. Similarly, the first position to the right of the decimal represents the digit multiplied by the numerical base taken to the power of negative one, the second position represents the digit multiplied the numerical base taken to the negative square power, and so on. The results are all added together, giving us the total value of the number. This is true of base-10, base-60, or any other numerical base.

Stated most generally, any number D can be represented in any base-N (N>0) by a series of digits d as follows:

$$D = d_j N^j + d_{j-1} N^{j-1} + ... + d_1 N^1 + d_0 N^0 . d_{-1} N^{-1} + d_{-2} N^{-2} + ... + d_{-k} N^{-k}$$, where all digits d satisfy the constraint $0 \leq d_i < N$.

In positional notation this is written:

$$D = (d_j d_{j-1} ... d_1 d_0 . d_{-1} d_{-2} ...)_N$$, where all digits d satisfy the constraint $0 \leq d_i < N$.

Take, for example, the decimal (base-10) number "103.95", which is shorthand for:

$$1 \times 10^2 + 0 \times 10^1 + 3 \times 10^0 + 9 \times 10^{-1} + 5 \times 10^{-2} = 100 + 0 + 3 + ^9/_{10} + ^5/_{100}$$

Such is the power of positional notation, we immediately grasp the precise value of 103.95 without having to resort to the above arithmetic.

S³: The Smith Sexagesimal System

Positional sexagesimal is just as intuitive with a little practice. In sexagesimal, each position to the left of the radix point represents increasing powers of 60, while those to the right of the radix point represent increasing negative powers of 60.

Consider "𝔈0/.𝞧𝟛". Each symbol represents a number between zero and fifty nine (sexagesimal symbols are defined and explained in detail in the next chapter). As in decimal, the position of the symbol defines the magnitude of its value. In the number above, 𝔈 represents 48, 0 remains 0, / is 1, the radix point remains unchanged, 𝞧 is 15, and 𝟛 is 39.

Sexagesimal positional notation yields:

𝔈 x /0² + 0 x /0¹ + / x /0⁰ + 𝞧 x /0⁻¹ + 𝟛 x /0⁻²

Written with equivalent but more familiar numerals:

$48 \times 60^2 + 0 \times 60^1 + 1 \times 60^0 + 15 \times 60^{-1} + 39 \times 60^{-2}$

This results in $172{,}800 + 0 + 1 + {}^{15}/_{60} + {}^{39}/_{3600} = 172{,}801 \; {}^{939}/_{3600}$, or the decimal value $172{,}801.26083\cdots$ (3 recurring).

Notice how a three digit numeral in base-60 can represent a six digit base-10 number. By requiring fewer digits, base-60 allows us to work with and retain much larger numbers than base-10 without having to resort to mnemonic tricks or memorization (which accesses long-term memory). However, when long-term memory is used, vastly larger values can be more easily memorized using sexagesimal.[8]

2 – WRITTEN SEXAGESIMAL NUMERALS

As in decimal, there is a unique numerical symbol for each sexagesimal digit. In base-10, these symbols are 0 through 9. In base-60, they are 0 through ৭ as shown in Figure 1.

Base-60 symbols consist of two parts, which are joined to create a sexagesimal numeral. The upper portion represents 0-9, the lower portion an additive value (+10, +20, up through +50). The uppermost row and leftmost column of the grid show each partial component, while the rest of the grid shows how those components are combined to create the numerals 0-৭ (0-59).[9]

S³: The Smith Sexagesimal System

		+0	+10	+20	+30	+40	+50
		"I"	"ϯ"	"ɔ"	"ⴷ"	"ʌ"	
0	"o"	0=0	၃=10	ဥ=20	ဒ=30	၉=40	၉=50
1	"/"	/=1	⸝=11	⸝=21	⸝=31	⸝=41	⸝=51
2	"⅃"	⅃=2	⊁=12	⸝=22	⸝=32	⸝=42	⸝=52
3	"ω"	ω=3	ѱ=13	Ѱ=23	Ѱ=33	Ѱ=43	Ѱ=53
4	"ᴧ"	ᴧ=4	⸝=14	⸝=24	⸝=34	⸝=44	⸝=54
5	"∇"	∇=5	⸝=15	⸝=25	⸝=35	⸝=45	⸝=55
6	"U"	U=6	⸝=16	⸝=26	⸝=36	⸝=46	⸝=56
7	"ㄴ"	ㄴ=7	⸝=17	⸝=27	⸝=37	⸝=47	⸝=57
8	"ထ"	ထ=8	၉=18	၉=28	၉=38	၉=48	၉=58
9	"ʊ"	ʊ=9	၉=19	၉=29	၉=39	၉=49	၉=59

Figure 1. Sexagesimal Numerals

3 – SEXAGESIMAL NOMENCLATURE

In base-10, numbers larger than ten are named in an inconsistent and haphazard way. We say "fourteen", but not "fiveteen", "forty" but not "twoty", "threety", or "fivety". Larger numbers are even worse. "Hundreds", "thousands", and "millions" are fairly consistent, but "milliards" (10^9 European) means nothing to an American. "One billion" means 10^9 in some countries, but 10^{12} to others, "one trillion" may mean 10^{12} or 10^{15}, and so on. In fact, the BBC often resorts to saying "million million" rather than "billion". Really huge numbers often have no names at all, with the occasional exception arbitrarily thrown in to sow a little extra confusion (a "googol" is 10^{100}, but what do you call 10^{99} or 10^{125}?).

In its purest form, sexagesimal removes all of decimal's warts and simplifies the process of naming large numbers.

NAMING SINGLE DIGIT NUMBERS:
0-৭ (0-59)

Pure Nomenclature assigns a monosyllabic name to each digit /-৭ (1-59), as shown in Figure 2. It is tighter and more concise than much of the old decimal-derived terminology, removes all of the inconsistencies of decimal, and shines when used to articulate large or verbose numbers.

Pure Nomenclature has the disadvantage of being unfamiliar to most people. Those wishing to use sexagesimal without first learning new names for ⊥ (7) and ◀-৭ (11-59) may use Transitional Nomenclature.

In **Transitional Nomenclature**, all digits from 0 to ৭ (0-59) are pronounced and spelled exactly as they are in decimal. Transitional Nomenclature helps ease the learning curve, allowing a smoother transition from decimal to sexagesimal.

S³: THE SMITH SEXAGESIMAL SYSTEM

	"1s"	"10s"	"20s"	"30s"	"40s"	"50s"
0	Zero (0|0)	Ten (⁊|10)	Ben (⅊|20)	Jen (Ƨ|30)	Ken (Ə|40)	Ven (ၐ|50)
1	One (/|1)	Tonn (𝟏|11)	Bonn (ɮ|21)	Jonn (ꝸ|31)	Konn (ɛ|41)	Vonn (ſ|51)
2	Two (⅃|2)	Toe (⊤|12)	Boe (ƃ|22)	Joe (ꝫ|32)	Koe (ȝ|42)	Voe (ㅈ|52)
3	Three (ɯ|3)	Tree (Ψ|13)	Bree (ꝮЬ|23)	Jree (Ꙍ|33)	Kree (ꝥ|43)	Vree (𐌙|53)
4	Four (Δ|4)	Tor (✝|14)	Bor (₿|24)	Jor (ꝩ|34)	Kor (ꝣ|44)	Vor (⋔|54)
5	Five (▽|5)	Tive (Ƴ|15)	Bive (ƻ|25)	Jive (ƽ|35)	Kive (Ƹ|45)	Vive (ꙗ|55)
6	Six (∪|6)	Tix (Ƴ|16)	Bix (ꝫ|26)	Jix (ɛ|36)	Kix (ꝫ|46)	Vix (ꝩ|56)
7	Sev (⌐|7)	Tev (✝|17)	Bev (ƫ|27)	Jev (ꝭ|37)	Kev (ꝭ|47)	Vev (ⱴ|57)
8	Eight (ಐ|8)	Taite (ꝑ|18)	Baite (ꝏ|28)	Jaite (ꝏ|38)	Kaite (ꝑ|48)	Vaite (ꝓ|58)
9	Nine (ᘔ|9)	Tine (ꝙ|19)	Bine (ꝕ|29)	Jine (ꝙ|39)	Kine (ꝑ|49)	Vine (ꝓ|59)

Figure 2. Pure Nomenclature for Sexagesimal Digits

NAMING MULTI-DIGIT NUMBERS

By applying five simple rules, the name of any sexagesimal number is easily derived, as is the name and prefix of its corresponding Metric-60 unit modifier (analogous to "milli" and "kilo" in Metric-10 nomenclature). A sixth rule defines how to abbreviate the Metric-60 prefix.

The Six Rules of Sexagesimal Nomenclature

Rule 1: Choose either Pure or Transitional Nomenclature:

If using Pure Nomenclature, write and pronounce each digit 0-𝟾 (0-59) as described in Figure 2.

If using Transitional Nomenclature, use the decimal name for each digit 0-𝟾 (0-59).

Rule 2: The base consonant defines the absolute value of the exponent (see Figure 3). "Absolute Value" simply means that in applying Rules Two and Three we ignore the sign and treat negative exponents exactly as we do positive exponents.

S³: The Smith Sexagesimal System

/0ʹ	60^1	b	("b" as in "Bravo")
/0ˀ	60^2	d	("d" as in "Delta")
/0ω	60^3	f	("f" as in "Foxtrot")
/0ᴧ	60^4	g	("g" as in "Golf")
/0∇	60^5	h	("h" as in "Hotel")
/0ᴗ	60^6	j	("j" as in "Juliette")
/0ᶦ	60^7	l	("l" as in "Lima")
/0ω	60^8	m	("m" as in "Mike")
/0ᵠ	60^9	n	("n" as in "Noon")
/0ˀ	60^{10}	p	("p" as in "Papa")
/0ᛉ	60^{11}	q	("q" as in "Quebec")
/0⁺	60^{12}	r	("r" as in "Romeo")
/0ᵠ	60^{13}	s	("s" as in "Sierra")
/0†	60^{14}	t	("t" as in "Tango")
/0ˠ	60^{15}	v	("v" as in "Victor")
/0ᵠ	60^{16}	w	("w" as in "Whiskey")
/0†	60^{17}	y	("y" as in "Yankee")
/0ᵠ	60^{18}	z	("z" as in "Zulu")
/0ᵠ	60^{19}	ch	("ch" as in "Channel")
/0ઠ	60^{20}	th	("th" as in "Thin")

Figure 3. Deriving the base consonant of a sexagesimal numerical prefix

+0	+0	a	("a" as in "c**a**t")
+⊱	+20	e	("e" as in "sh**e**d")
+⊰	+40	i	("i" as in "s**i**t")
+/0	+60	o	("o" as in "sh**o**w")
+/⊱	+80	u	("u" as in "Z**u**lu")
+/⊰	+100	y	("y" as in "sp**y**")
+⟍0	+120	æ	("a" as in "g**a**te")
+⟍⊱	+140	ē	("e" as in "sh**ee**t")

Figure 4. Deriving the base vowel of a sexagesimal numerical prefix

Rule 3: The base vowel modifies the consonant, adding 20, 40, 60, 80, 100, 120, or 140 to the exponent (see Figure 4).

Rule 4: The prefix "an" is prepended for negative exponents.

Rule 5: Add "zend" as the final syllable for base-60 numbers. Add "ra" as the final syllable for Metric-60 units.

Rule 6: For Metric-60 units the base consonant and base vowel define the unit modifier abbreviation (analogous to k for kilo, m for milli, etc.), which is separated from the unit name or abbreviation by a hyphen. For example, meratocks are abbreviated me-t. Negative exponents prefix the abbreviation with "an". For example, anbarasparks (a very small electrical charge) would be abbreviated anba-s.

S³: The Smith Sexagesimal System

Deriving Sexagesimal Nomenclature by Example

Example 1 – "a bazend":

Consider /0 (60):

Rule 1: Selecting Transitional Nomenclature tells us that we pronounce numerals 0-⅞ (0-59) as we would in decimal. /0 (60) falls outside of this range.

Rule 2: Our first non-zero digit is in the second place to the left of the radix point, the /0′ (60^1) position. The absolute value of the exponent of /0′ (60^1) is / (1). Applying Rule Two, we get "b" as our base consonant. (See Figure 3.)

Rule 3: We do not need to add anything to the exponent, so +0 (+0) gives us "a" as our base vowel. (See Figure 4.)

Rule 4: Our exponent is a positive value, so Rule Four does not apply.

Rule 5: Since we are discussing a number, and not a Metric-60 unit modifier, we append "zend" to our base syllable, giving us the word "bazend".

Thus, /0 is written and pronounced "one bazend" or "a bazend", never "sixty".

Example 2 – A Simple Number:

Consider /Ы (106). This number is written and pronounced "one bazend forty-six" in Transitional Nomenclature. See if you can apply the five rules above to come up with this result. If you can, you are well on your way to understanding sexagesimal.

Example 3 – Fractions and Negative Exponents:

Consider a temperature of 0.0ധ⅞ degrees:

Rule 1: Selecting Transitional Nomenclature, Rule One tells us to pronounce each digit as we would its decimal equivalent.

Rule 2: The first non-zero numeral is two places to the right of the radix point, in the /0⁻ᴧ position (60⁻²), so our exponent is -ᴧ (-2). The absolute value of -ᴧ (-2) is ᴧ (2) (just drop the minus sign). Applying Rule Two, our base consonant is "d". (See Figure 3.)

Rule 3: We do not need to add anything to the exponent, so +0 (+0) gives us a base vowel of "a". (See Figure 4.)

Rule 4: We are dealing with a negative exponent, so we prefix our base stem with "an", giving us "anda".

Rule 5: If we wish to state the value as a number, 0.0ய𐤒 degrees would be pronounced either "zero point zero three fifty-eight degrees" or "three andazend, fifty-eight anfazend degrees". If we wish to state the value in terms of Metric-60 units, we have "three point fifty-eight andaradegrees".

Rule 6: If the value is written in terms of Metric-60 units, we abbreviate ய.𐤒 andaradegrees as ய.𐤒 anda-° (or ய.𐤒 anda-deg).

Example 4 – Really Big Numbers:

Consider a very long distance: a bazend to the bazend nineteen meters (/0^/𐤒 m, or 60⁷⁹ m):

Rule 1: To keep things simple, we'll stick with Transitional Nomenclature.

Rule 2: We note that /𐤒=𐤒+/0 (79=19+60). Applying Rule Two, 𐤒 (19) yields a base consonant of "ch". (See Figure 3.)

Rule 3: To get an exponent of /𝕈 (79) we need to add /0 (60) to the value 𝕈 (19) used in Rule Two to derive our base consonant. Therefore, applying Rule Three, +/0 (+60) gives us "o" is our base consonant, yielding "cho" as our stem. (See Figure 4.)

Rule 4: We are working with a positive exponent, so Rule Four does not apply.

Rule 5: According to Rule Five, we append "zend" if we are naming a number, and "ra" if we are referring to a Metric-60 prefix. So, we have one chozend meters (/0$^{/𝕈}$ or 60^{79} meters) or one chorameter (abbreviated / cho-m according to Rule 6). How big a distance is that? Further than the limits of the currently observable universe, and further than those limits are likely to expand during the lifetime of the universe. In base-10, a chorameter is about $2.97816517 \times 10^{140}$ meters, or the distance light will travel in $3.14788031 \times 10^{124}$ years. How much time is that? It's about 10^{114} times longer than the universe is old.

Example 5: More Complex Numbers:

Consider 𝔈𝕐𝟛,𝔷𝔚𝕩.𝕈𝕐 (37,527,784,196 .179444⋯):

𝔈𝕐𝟛,𝔷𝔚𝕩.𝕈𝕐 = 𝔈 x /0$^{∇}$ + 𝕐 x /0$^{∆}$ + 𝟛 x /0^{ω} + 𝔷 x /0$^{∆}$ + 𝔚 x /0$^{/}$ + 𝕩 x /0^{0} + 𝕈 x /0$^{-/}$ + 𝕐 x /0$^{-∆}$ (48 x 60^5 +15 x 60^4 +39 x 60^3 + 44 x 60^2 + 29 x 60^1 + 56 x 60^0 + 10 x 60^{-1} + 46 x 60^{-2})

Rule 1: We'll continue to use Transitional Nomenclature.

Rule 2: Our first non-zero digit is 𝔈 x /0$^{∇}$ (48 x 60^5). The absolute value of the exponent is ∇ (5), giving us a base consonant of "h". (See Figure 3.)

Rule 3: We do not need to add anything to our exponent, so "a" is our base vowel. (See Figure 4.)

Rule 4: The exponent is positive, so Rule Four does not apply.

Rule 5: Our stem is "ha". ℰ x /0$^\triangledown$ (48 x 60^5) is thus "forty-eight hazend".

Iterating through the five rules for our remaining digits: Ƴ (15), has an exponent of ∆ (4), giving us "ga" as our stem, yielding "fifteen gazend". ꙛ (39) has an exponent of ω (3), giving us "fa" as our stem and "thirty-nine fazend" as our number. ⱬ (44) has an exponent of ⅄ (2), yielding "forty-four dazend", while ꙙ (29) has an exponent of / (1), giving us "twenty nine bazend". Ⴘ (56) is in the units position, and thus is simply pronounced "fifty six".

To the right of the radix point we have ꟼ (10) with an exponent of -/ (-1), giving us "ten anbazend", followed by ⱨ (46) times /0$^{-\lambda}$ (60^{-2}), thus having an exponent of -⅄ (-2) and yielding "forty-six anfazend".

In Transitional Nomenclature, the entire number is therefore pronounced "forty-eight hazend, fifteen gazend, thirty-nine fazend, forty-four dazend, twenty-nine bazend, fifty-six point ten anbazend, forty-six andazend". This can be shortened to "forty-eight hazend, fifteen ga, thirty-nine fa, forty-four da, twenty-nine ba, fifty-six point ten anba forty-six andazend".

Example 6 – Short Form:

When Short Form is used, the full name is only applied to the first digit before the radix point and the last digit after the radix point. All other syllables use the abbreviated prefix only, and any zero digit is ignored.

For a simple example, consider ℨᏫᎰᎩ (9,768,619). Using Transitional Nomenclature, in full form we would say "forty-five fazend, thirteen dazend, thirty bazend, nineteen". However, in Short Form we would say "forty-five fazend, thirteen da, thirty ba, nineteen".

Example 7 – A Complex Number (Short Form):

Consider ⇌0,℘0⇌,0∇/.0℘0⍙
(123,209,170,632,301.0119484568):

Rule 1: Well do this both ways, first using Transitional Nomenclature, and then compare the result with the same value using Pure Nomenclature.

Rules 2-4: Our first exponent is ∠ (7), so using Transitional Nomenclature, Rules One through Four give us "forty-four lazend" as our full numerical name for this digit. In both Long Form and Short Form we don't bother with zero (0) digits, so our next nonzero digit is ℘ (48), which has an exponent of ∇ (5), yielding "forty-eight hazend". Our next non-zero digit is ⇌ (27) with an exponent of ω (3), giving us "twenty-seven fazend", followed by "zero dazend" (which is dropped), "five bazend", "one", "zero anbazend", "forty-three andazend", "zero anfazend", "fifty-two angazend".

Thus, ⇌0,℘0⇌,0∇/.0℘0⍙
(123,209,170,632,301.0119484568) is spoken and written as "forty-four lazend, forty-eight hazend, twenty-seven fazend, five bazend one point forty-three andazend fifty-two angazend" (Long Form) or "forty-four lazend, forty-eight ha,

twenty-seven fa, five ba, one point forty-three anda, fifty-two angazend" (Short Form).

Using Pure Nomenclature, ₴0,₰0₺,0▽/.0₽0⸓ (123,209,170,632,301.0119484568) is spoken and written as "kor lazend, kaite hazend, bev fazend, five bazend one point kree andazend voe angazend" (Long Form) or "kor lazend, kaite ha, bev fa, five ba, one point kree anda, voe angazend" (Short Form). This conciseness when dealing with large or verbose numbers is a powerful incentive for learning and using Pure Nomenclature.

4 – CONVERTING BETWEEN SEXAGESIMAL AND DECIMAL

The algorithm for converting between any two bases N and M, where N>0, M>0, and M≠N is quite straightforward. Recall from our introduction that any integer and fraction in base-N (N>0) can be represented as:

$D = (d_j d_{j-1} ... d_1 d_0 . d_{-1} d_{-2} ...)_N$, where all coefficients d_i are constrained by $0 \leq d_i < N$.

Two algorithms are required to convert between bases: one algorithm for the integer portion of the number (to the left of the radix point), and another for the fraction (to the right of the radix point).

First, consider the integer portion of the number, I. Set exponent $E = 1$. Divide I by N^E. Is the integer portion of the result less than N? If not, increment E by 1 and divide I by

N^E. Continue until the integer portion of the result is less than N. Once the integer portion of the result is less than N, we have our first digit d_j. Derive $I_j = I - d_j N^E$ and divide the result by N^{E-1}. This yields our second digit, d_{j-1}. Calculate $I_{j-1} = I_j - d_{j-1} N^{E-1}$. Continue deriving digits d_i from I_i in this fashion, decrementing the exponent until $I_i < N$. At this point we have d_0 and are finished.

Now, consider the fraction. Multiply the fraction by N. The integer portion of the result is our first digit to the right of the radix point. Drop the integer portion, and multiply the remaining fraction by N again. The integer portion of the result yields our next digit. Continue until you calculate the last digit, discover it is recurring, or reach the desired level of precision.

This algorithm is best shown by example. Consider the decimal number 6,431.275. We wish to convert this number to sexagesimal.

First, we consider the integer portion of the number: 6,431.

$^{6431}/_{60} = 107.183333$. Is 107 < 60? No, so increment the exponent.

$^{6431}/_{3600} = 1.7863889$. Is 1 < 60? Yes. Our first digit is **/** (1).

6431 − 1 x 3600 = 6431 − 3600 = 2831

$^{2831}/_{60} = 47.1833333$. Our second digit is **č** (47).

2831 − 47 x 60 = 11. **ᚒ** (11) is less than 60, so it is our third and last digit. The integer portion of our number is (1)(47)(11), or **/čᚒ**.

Now consider the fractional portion of the number: 0.275.

0.275 x 60 = 16.5. Our first digit to the right of the radix point is Ynn (16).

Drop the 16 and consider the remaining fraction: 0.5

0.5 x 60 = 30.0 Our second digit to the right of the radix point is ϟ (30). Since the remaining fraction is zero, the sequence has converged and we are done. The fractional portion of our number is (16)(30), or Ynnϟ.

Thus, we have converted 6431.275_{10} to $(1)(47)(11).(16)(30)_{60}$, or "/ϟ⸓⸗.Ynnϟ".

PART 2

NATURAL METRIC-60 UNITS OF MEASURE

5 – MAXWELL PLANCK'S "NATURAL UNITS" AND METRIC-60 DEFINED

In the Planck scale (and corresponding Metric-60), units associated with fundamental physical theory, such as c (the speed of light in a vacuum), $\hbar$ (Dirac's constant), G (the gravitational constant), k_c (the Coulomb constant), and k_B (the Boltzmann constant), are all defined as equal to one. Planck units thus derive not from arbitrary human convention, but from a fundamental, physical level. Relationships that are sometimes non-intuitive become obvious. For example, $e=mc^2$ becomes $e=m$ (energy equals mass), $E=\hbar w$ becomes $E=w$ (energy equals angular frequency), and so on.

Slight uncertainties exist in the measurement of the gravitational constant upon which Planck values depend. Because Metric-60 is based on Planck units, it suffers this same limitation. However, when the value of the gravitational constant is refined, Metric-60 will facilitate an orderly migration to a new, more precise set of units. Metric-60 units have different names than their Planck counterparts, so adopting new units with new names is straightforward. Figure 5 shows an overview of the four fundamental units of Metric-60, the Planck units from which they derive, their dimensionality, and their SI equivalents.

Metric-60 Unit	Planck Unit	Dimension	SI equivalent
Degrees (Range: 60^{19} °P)	Planck Temperature	Temperature (ML^2T^{-2}/k)	1.416833×10^{32} K
Tock (t)	Planck Time Planck Length	Time (T) Length (L)	5.39106×10^{-44} s 1.616199×10^{-35} m
Speck (s)	Planck Mass	Mass (M) Energy (E)	2.17651×10^{-5} g 1.956×10^9 J
Spark (e)	Planck Charge	Electric Charge (Q)	$1.875545956 \times 10^{-18}$ C

Figure 5. Metric-60 Base Units, Planck Base Units, and their SI equivalents.

6 – PLANCK TEMPERATURE AND DEGREES PRIME

The Planck Temperature is the temperature one Planck Time unit after the big bang. According to current theory, no physically meaningful temperature hotter than this exists. The possible range of any temperature anywhere in the physical universe, at any time from its beginning to its end, is bounded by absolute zero (the coldest possible temperature) and the Planck Temperature (the hottest possible temperature).

I have divided this range into an arbitrary scale of 60^{19} degrees "Prime", resulting in a conversion factor of 0.0232511750772 °K/°P, or 43.0098843 °P/°K. Because later theories might yield different values for the Planck temperature, new scales might need to be adopted. For simplicity, future Metric-60 scales would be named after the Latin ordinals "Secondus", "Tertius", "Quatrus", and so on.

	Prime	**Kelvin**	**Celsius**	**Fahrenheit**
Absolute Zero	0° P (0)	0	-273.15°	-459.67°
bitter cold	ⱳⱳⱳ° P (10,983)	255.37	-17.78°	0°
H_2O Freezing	ⱳ𐒠0° P (11,748)	273.15	0°	32°
chilly	ⱳ𐒧𐒠° P (12,178)	283.15	10°	50°
pleasant	ⱳ𐒉𐒎° P (12,608)	293.15	20°	68°
perfect	ⱳ𐒐𐒤° P (12,704)	295.37	22.22°	72°
hot	ⱳ𐒐𐒠° P (13,038)	303.15	30°	86°
very hot	ⱳ𐒧𐒳° P (13,373)	310.93	37.78°	100°
H_2O Boiling	𐒃𐒧𐒤° P (16,049)	373.15	100°	212°
Lead Melts	𐒃𐒠𐒳° P (25,813)	600.16	327.46°	621.43°
Sun's Surface	/ⱷ𐒰𐒐° P (249,457)	5800	5528.85°	9980.33°
Atomic Bomb	𐒠𐒐ⱷ𐒤° P (12,902,965)	300000	299,727°	~540,000°
Sun's Core	𐒣.𐒤𐒠ⱷ𐒠 ga° P (51.77)	15,600,000	15,599,727°	28,079,540°
Planck Temp.	/ cha° P (1)	1.416785 x 10^{32}	1.416785 x 10^{32}°	2.55 x 10^{32}°

Figure 6. Common Temperatures

Figure 6 shows several common and familiar temperatures in degrees Prime, Kelvin, Celsius, and Fahrenheit. While the temperatures in degrees Prime seem large to those used to decimal numbers, keep in mind that in sexagesimal these values are mostly three and four digit temperatures. For example, ⲱⲱⲱ° P (10,983° P) is 0° F or -18° C, while /ⲚϤϪ° P (249,456° P) is the temperature of the sun's surface (5529° C). For those used to sexagesimal, these numbers are as easy to remember as their Kelvin, Celsius, or Fahrenheit equivalents, and at higher temperatures, significantly easier.

7 – PLANCK TIME AND "TOCKS"

Planck Time, denoted as t_p, is the smallest measurement of time that has any meaning (any events occurring closer together are simultaneous). t_p is defined formally as the time it would take a photon traveling at the speed of light to cross a distance equal to the Planck length, and is approximately 5.39106×10^{-44} seconds. The Metric-60 unit "tock" (t) is derived from physicists' best estimation of the Planck Time, and is defined as precisely 5.39106×10^{-44} seconds. Figure 7 shows some common durations of time and their Metric-60 equivalents.

Natural Metric-60 Units of Measure

Description	Metric-60 Units	SI Units
Planck Time	∕ t (1)	5.39121×10^{-44} s
Second	⏶.⏴⏴⏵⏷⏶ ge-t (3.91457)	1 s
Minute	⏶.⏴⏴⏵⏷⏶ he-t (3.91457)	60 s
Gen-5 Circadian*	⏶.⏴⏵⏶⏷⏶ he-t (4.69748)	72 s
Gen-4 Circadian*	⏶.⏵⏷⏶⏷⏶ he-t (5.63698)	86.4 s
Gen-3 Circadian*	⏶.⏷⏴⏶⏷⏶ he-t (9.39496)	144 s
Gen-2 Circadian*	⏶.⏶⏷⏶⏶⏶ he-t (28.18488)	432 s
Gen-1 Circadian*	⏶.⏶⏷⏶⏷0 je-t (3.13165)	2880 s
Hour	⏶.⏴⏴⏵⏷⏶ je-t (3.91457)	3600 s
Day	∕.⏷⏶⏷⏶⏶ le-t (1.56583)	86,400 s
Week	⏶.⏴⏶⏶⏶⏴ le-t (10.96079)	604,800 s
Year	⏶.⏶⏴⏶⏷⏴ me-t (9.53177)	31,556,926 s
Approximate Age of the Earth	⏶.⏶⏶⏶⏶⏴ se-t (55.16069)	4.5×10^{9} years
Approximate Age of the Universe	⏶.⏶⏷⏶⏶0 te-t (2.7989)	1.37×10^{10} years

Figure 7. Common Durations of Time in Metric-60 and SI Units.
*Fictional measure of subjective time from the novel *Autonomy*

8 – PLANCK LENGTH AND "TOCKS"

In addition to being the fundamental unit of time, the tock (t) is also the fundamental measure of distance. It is defined to be equal to the Planck Length. The Planck Length, l_p, is the smallest distance that has any meaning in the physical universe. Below this distance quantum mechanics makes any and all measurements nonsense. It is the maximum resolution of space and time in our physical universe. Because time and space are unified, one tock in distance is exactly the distance light travels, in a vacuum, in one tock of time (1.616199 x 10^{-35} meters). Figure 8 shows some common distances. For a sense of scale, the term qaratock (qa-t) refers to the femto scale, saratocks (sa-t) to the atomic scale, and taratocks (ta-t) to the nano scale. Yaratocks (ya-t) are roughly analogous to millimeters, zaratocks (za-t) to centimeters or inches, charatocks (cha-t) to feet or meters, beratocks (be-t) to kilometers or miles, and meratocks (me-t) to light-years. Note

Natural Metric-60 Units of Measure

the time=space relationship. Meratocks are also analogous to years when measuring time. This is not a coincidence.

	Metric-60 Units	**SI Units**
Planck Length	/ (1) tock	1.616199×10^{-35} m
Radius of a Proton	~/.ਣ੧੪₺ ga-t (~1.7054)	~1×10^{-15} m
Atomic radius: hydrogen	~੪.੨੩੩₺ sa-t (~56.8472)	~1.2×10^{-10} m
Millimeter	੪.੪੧੨ ya-t (36.553)	0.001 m
Centimeter	౦.∇᠐੧⊦ za-t (6.0922)	0.01 m
Inch	੧.੪੪₺ za-t (15.474)	0.0254 m
Foot	ω.∇ᘓ੮੨ cha-t (3.0948)	0.3048 m
Meter	੧.౦⊦↑੪ cha-t (10.1536)	1 m
Kilometer	⅃.੨੪੪₺ be-t (2.8204)	1000 m
Mile	⅃.ᘓ੪੪੪ be-t (4.5391)	1609.344 m
Astronomical Unit	ᘓ.੪੪౦੪ he-t (32.5566)	149,598,000,000 m
Light Year	౦.ᘓ⊦੪੨ me-t (9.5318)	9.4605284×10^{15} m

Figure 8. Common distances in Metric-60 and SI Units.

9 – PLANCK MASS AND "SPECKS"

The Planck mass is the mass at which a body's Compton length (the distance at which quantum mechanical properties dominate) and the Schwarzschild radius (the radius at which, if a mass is squeezed down to that size, it becomes a black hole) are equal. Put another way, the Planck mass is the mass at which both relativistic and quantum properties are equally dominant.

The Metric-60 unit for mass and energy is the speck (s), where a speck is defined as equal to one Planck mass, or 2.17651×10^{-5} g.

10 – PLANCK CHARGE AND "SPARKS"

The Planck charge, q_p, is defined in terms of the speed of light (c), Planck's constant (h), and the permittivity of free space (ε_0):

$$q_p = (2ch\varepsilon_0)^{1/2}$$

The Metric-60 unit for charge is the spark (e), where one spark is defined as equal to the Planck charge, or $1.875545956 \times 10^{-18}$ C. This is about 11.7 times the charge of an electron (ignoring the sign).

11 – DERIVED UNITS

From the four Metric-60 base units—tocks, specks, sparks, and degrees Prime—all other units are derived, whether for volume, momentum, energy, force, density, frequency, pressure, impedance, or any other physical measure. Often, dimensions (such as mass and energy or time and distance) will cancel each other out, so dimensionality of these measures is important. Some examples are shown in Figure 9.

Natural Metric-60 Units of Measure

	Metric-60 Units	SI Units
Planck Area (L^2)	/ t^2 (1 tock^2)	2.61223×10^{-70} m^2
Planck Volume (L^3)	/ t^3 (1 tock^3)	4.22419×10^{-105} m^3
Planck Momentum (LMT^{-1})	/ s (1 speck)	6.52485 kg m/s
Planck Energy (L^2MT^{-2})	/ s (1 speck)	1.9561×10^9 J
Planck Force (LMT^{-2})	/ s/t (1 speck/tock)	1.21027×10^{44} N
Planck Power (L^2MT^{-3})	/ s/t (1 speck/tock)	3.62831×10^{52} W
Planck Density ($L^{-3}M$)	/ s/t^3 (1 speck/tock^3)	5.155×10^{96} kg/m^3
Planck Angular Frequency (T^{-1})	/ t^{-1} (1 tock^{-1})	1.85487×10^{43} s^{-1}
Planck Current (QT^{-1})	/ e/t (1 spark/tock)	3.4789×10^{25} A
Planck Voltage ($L^3MT^{-2}Q^{-1}$)	/ ts/e (1 tock-speck/spark)	1.04295×10^{27} V
Planck Impedance ($L^2MT^{-1}Q^{-2}$)	/ ts/e^2 (1 tock-speck/spark^2)	29.9792458 Ω
Planck Pressure ($L^{-1}MT^{-2}$)	/ s/t (1 speck/tock)	4.63309×10^{113} Pa

Figure 9. Commonly Derived Units in Metric-60 and SI Units.

12 – FICTIONAL METRIC-60 UNITS FROM THE NOVEL *AUTONOMY*

SUBJECTIVE TIME IN THE VIRTUAL

In my novel *Autonomy* characters experience time at different rates depending on whether they are operating in the Virtual or out in the physical world. Subjective time in a virtual environment is a function of the computational complexity of the mind running as software, the software environment it is operating in, and the speed of the hardware it is running on. As a result, the relationship between subjective time in the Virtual versus objective time in the Physical changes as software is optimized and new hardware deployed.

To reconcile such complex temporal relationships the Community adopts a unique unit for measuring subjective time: the circadian. One circadian is defined as precisely 24 hours of subjective time. Early in the novel the Community initially divides circadians into Metric-10 units, but later, when they adopt the sexagesimal numeral system, they also standardize on a Metric-60 System that includes circadians divided into Metric-60 units.

Thus, one circadian is made up of sixty anbaracircadians (anba-C), which in turn are made up of sixty andaracircadians (anda-C), each of which is made up of sixty anfaracircadians (anfa-C). These are analogous to virtual (but short) hours, minutes, and seconds respectively. Likewise, sixty circadians are equal to a baracircadian (ba-C), which is analogous to a virtual month, while sixty baracicadians make up a daracircadian (da-C), which is analogous to a virtual decade.

Natural Metric-60 Units of Measure

Description	Metric-60 Units	SI Units
Planck Time	⁄ t (1)	5.39121×10^{-44} s
Second	⍵.✦✶𐊠⌿✸ ge-t (3.91457)	1 s
Minute	⍵.✦✶𐊠⌿✸ he-t (3.91457)	60 s
Gen-5 Circadian	⌂.≥𐊨𐊠ℰℙ he-t (4.69748)	72 s
Gen-4 Circadian	∇.✸✤⌿ℰℙ he-t (5.63698)	86.4 s
Gen-3 Circadian	⌀.✇ℰ↑↕✵ he-t (9.39496)	144 s
Gen-2 Circadian	ℙ.↑∇ℒ⌂ℙ he-t (28.18488)	432 s
Gen-1 Circadian	⍵.⌿✤✵⊕0 je-t (3.13165)	2880 s
Hour	⍵.✦✶𐊠⌿✸ je-t (3.91457)	3600 s
Day	⁄.✵✵𐊨✵ℙ le-t (1.56583)	86,400 s

Figure 10. Common Durations of Subjective and Objective Time.

DATES AND OBJECTIVE TIME IN THE PHYSICAL

In *Autonomy*, objective time is measured against physical reality in meratocks. Since no two people experience exactly the same rate of subjective time, dates are also expressed against objective time, in meratocks, measured from an arbitrary "zero-point" in time. In *Autonomy*, the zero-point is defined as the moment the first human onloads into the Virtual.

Metric-60 dates are written as follows:

m.jj-h:g:f.x meratocks (me-t) New Epoch

where m=meratocks, jj=jeratocks, h=heratocks, g=geratocks, f=feratocks, and x is any remaining fraction.

The easiest way to come up with a given Metric-60 date in terms of "tocks" is to figure out how many seconds have passed since time 0, convert the result to feratocks using the conversion factor of 234.874026 fa-t/s, derive the resulting sexagesimal numeral, and then insert the date-time punctuation into the resulting integer.

For example, in *Autonomy* the first human onload takes place at:

0.00-0:0:0 me-t New Epoch.

The Autonomous Community's worst crisis comes to a head about 8,311,507 seconds later. 8,311,507 seconds x 234.874026 feratocks/second yields 1,952,157,110 feratocks. Converting this number to sexagesimal using the procedure described in Chapter 4 results in the value: ⟑𝟛𝟜𝟝𝟙𝟠.

NATURAL METRIC-60 UNITS OF MEASURE

Inserting this value into the standard time-date punctuation above yields the date expressed in the following form:

⊻.ϑϚ-ϖ:ϟ:ϙ me-t new Epoch

NOTES

1 Farrington, Jeanne (2011). "Seven plus or minus two". *Performance Improvement Quarterly 23* (4): 113–6.

2 Cowan, Nelson (2001). "The magical number 4 in short-term memory: A reconsideration of mental storage capacity". *Behavioral and Brain Sciences 24* (1): 87–114; discussion 114–85.

3 Duncan J. Melville, *Mesopotamian Mathematics*, http://it.stlawu.edu/~dmelvill/mesomath/Numbers.html

4 *Ibid.*

5 *Ibid.*

6 *Ibid.*

7 There are many other ways to represent larger numbers like this, some even more difficult to read. See for example http://www.ancientscripts.com/sumerian.html

8 Stanislas Dehaene (1997). *The Number Sense: How the Mind Creates Mathematics*: 131-4.

9 Sexagesimal TrueType fonts are available free for non-commercial use under the Creative Commons Attribution-ShareAlike 4.0 International License, from http://autonomyseries.com/

A preview

of

AUTONOMY

a Novel

by

Jean-Michel Smith

AUTONOMY

Jean-Michel Smith

TIME-LAPSE

Tuesday, July 17, 2068
*1st **circadian** (0 seconds elapsed)*

Cal opened his eyes and sat up. The bed was large and decadently soft, surrounded by gauzy curtains hanging from a canopy through which shafts of golden sunlight shone.

"Onload complete, guys. It worked."

He pushed the curtain aside and jumped to his feet. Hilltop meadows surrounded him, lush green pastures sporting constellations of blue and violet flowers. He relished the feel of soft grass between his toes.

"The simulation is fantastic. Perfect weather, and what a view!"

To the east rose a spectacular range of mountains, snow covered slopes textured with stone and ice climbing to dramatic, pointed summits. Softened by haze, a massive planet peaked over their ridges, its Jovian nature betrayed by its green and gold swirls. To the west, in the distance, a sea reflected afternoon sunlight.

"Something's wrong with the light diffusion. The haze along the horizon isn't consistent. Not a big deal, though. Amazing!" A flock of pelicans flew past. Cal laughed. "Hey, who came up the pink birds?"

His grin faded.

"Doctor Nolen? Céline? Can you hear me? Acknowledge please."

He waited. Another flock of pelicans passed overhead.

"Node. Command mode engage."

A soft, neutral voice spoke within his mind. »Command mode engaged.«

Cal thought furiously. There could be a communications glitch. That was actually more likely than a systems malfunction at this point. Still, this was all very experimental. He'd better err on the side of caution.

"Run test suite one, systems integrity check," he commanded.

»Running ... Suite one complete. All operating parameters nominal.«

"Run suite two."

»Running ... Suite two complete. All operating parameters nominal.«

Cal forced himself to remain calm. They would bring him out after ten minutes no matter what.

"Run the third test suite."

»Running ... Suite three complete. All operating parameters nominal.«

"How long since I onloaded?"

»Time elapsed: two minutes, fifteen seconds.«

Cal paced back and forth along the crest of the hill. "Run a diagnostic on the external comm link."

»Running ... Initial protocol state achieved. Ping tests commencing.«

Long minutes of silence did little to calm Cal's jittery nerves. "You should have some results by now. What's taking so long?"

»Communications diagnostic still running. No errors detected.«

"Then why aren't they answering?"

»Insufficient data.«

His unease grew, his critical eye finding numerous details in the simulation that were not quite right, from the fractal fuzziness at the limits of his vision when he examined the grass, to the two dimensional quality of the clouds moving across the sky. "Damn it, we should have some kind of communication by now." His dread grew to outright fear. "How much time has elapsed since I onloaded?"

»Fourteen minutes, twenty-nine seconds.«

Cal stopped. "Say again?"

»Fourteen minutes, thirty-one seconds.«

Why was he still here? Why hadn't the offload sequence run as scheduled? He pictured his comatose body lying in the lab. He'd been crazy to volunteer for this experiment.

"What's the status of the communications check?"

»Link protocol is experiencing timing synchronization errors. No ping responses received.«

"Shit! Shit, shit, shit!"

Cal sank to the ground and put his face in his hands. Without a working data link there was no exit. He could be trapped, stuck in a software simulation, caught like a fly in amber in a world whose realism grew more fragile with each moment. Artificial comas were tricky things. When he didn't wake up they would try everything to bring him out of it. Eventually they might get desperate and reboot the equipment.

Cal's chest tightened into a knot. They'd been so enthusiastic, so carried away with their work, none of them had taken the idea seriously that he could get stuck in here. If they had, they would have included a lot more failsafes in the design, data persistence chief among them. If Cal survived that would be the first thing he'd correct. As it was, any kind of system reset would erase him. He would be stuck in a coma for the rest of his life, his physical brain as dead as his electronic self.

o—•

"What's happening?" Doctor Nolen sounded nervous. He cracked the door of the lab and scanned the empty hallway. Céline had been right, thank god. Ten thirty on a Tuesday morning in July was the best time to be doing this. The place was deserted.

"Well?" Doctor Nolen asked again, pulling the door shut and checking the lock.

Céline looked up from Cal's prostrate form. Thick cables sprouted from the metal lattice and electrodes surrounding his face and head. They braided their way across the floor, some to a refrigerator-sized bank of components wrapped in tubes of liquid nitrogen, others to medical monitors that showed a perfectly steady heartbeat.

Céline gripped the side of the steel table. "No telemetry, no audio, no video. Nothing."

Doctor Nolen hurried over to a bank of screens. "His coma is stable. Brain scan's a little quieter than I'd like. Did the onload sequence complete?"

"Over a minute ago." Her frown deepened. "Jesus! Why aren't we hearing from him?"

Footsteps clattered on the linoleum outside. Laughter echoed down the hall. Céline and Doctor Nolen stared at each other, hardly daring to breath until the noise faded away

"It's been nearly two minutes now," Céline whispered. "Something is very wrong."

⎯○

Cal's eyes traced the contours of the hills as they descended from the meadow toward the sea, forested hillsides plunging toward white sandy beaches and glittering turquoise water.

Without realtime communication he couldn't talk to Céline or Doctor Nolen directly, but maybe he could leave a trail of crumbs big enough for them to follow. Somewhere obvious. Somewhere they were bound to look once they knew there was a problem, like the buffer holding Céline's log files and software diagnostics.

"Node, record a message into the firmware log buffer."

»Persistent storage online. Ready to record.«

Cal took a deep breath to steady his voice. "Doctor Nolen, Céline. I'm alive and fully aware. Bring me back online. Do not reset the Node or wipe the software. We have two problems.

First, the communications link is down. Second, my automatic offload has failed. I suspect the two are related.

"I've been here almost an hour. I'll keep trying to establish contact. I've run the first three test suites successfully. I'm also running a diagnostic on the communications link, but it's taking far longer than expected. There's some kind of timing or synchronization problem with the protocol."

Cal stopped. An outrageous idea pushed its way to the front of his mind. "I might know what's wrong," he murmured, scrambling to his feet. " Guys, I'll get back to you! Node, end recording."

He needed to regroup, to think. "Teleport me to the coast."

The roar of the surf greeted him, the meadow around him replaced by a pristine beach of white sand. Could internal subjective time be running at a different rate than in the external world? He was shocked at the idea, but couldn't help feeling a little foolish for not thinking of it sooner.

Trying to calm himself, he sat beneath a wide palm and leaned against its trunk. "Node, how is signal synchronization defined?"

»Standard IPv12 protocol, synchronization timestamps based upon internal clock ticks.«

His mind raced. "Go into debug mode. Create a flat 2-d display at eye level in front of me and show me the code."

Forty minutes later Cal was still studying the communications protocol. A bell chimed.

»Diagnostic complete. Communications hardware OK. Protocol unable to synchronize with remote host. All signals have timed out.«

"Show me the current time-out settings."

A second display appeared in front of him. "5 milliseconds," Cal muttered. "That's a reasonable length of time. Node, there should be an external hardware clock available. Can you access it?"

»Yes.«

Cal felt the release of tension inside him. With an external timing source, he thought he could likely fix the problem. "OK. Node, measure the elapsing time on the hardware clock against that of the internal software clock. Compare and report." He stood up and walked toward the water.

»The internal clock is counting 30017 microseconds for each millisecond registered on the external clock.«

"So the time I'm experiencing in here is thirty times longer than that in the physical world?"

»Affirmative.«

"Wow! No wonder I didn't offload after ten minutes—it was only 20 seconds or so back in the lab."

Cal was shocked. A speedup of thirty was massive. If he was going to have any chance of fixing the data link, restoring communications and offloading back into the physical world, he'd need to clearly differentiate between real and subjective time.

"Node, define an internal clock with the following units: one 'circadian' equals one subjective 24-hour period of time, as measured by the internal software clock. Divide and multiply that unit using standard metric nomenclature." Cal's heart pounded. "Recode and calibrate all external communications protocols, referencing the objective clock and converting units as required. Confirm when finished."

»Modification successful.«

"Thank god!" Enormous relief left him feeling lightheaded. "How long will it take to rerun the diagnostics?"

»Full communications diagnostics will require approximately thirty-one point two five millicircadians, or precisely ninety seconds.«

Feeling more confident than he had since the experiment started, Cal waded out into the waves. Since he was here, he might as well give the environ's simulation software a workout.

"OK, run the diagnostics. Let me know when it's finished."

He dove underwater, swam several strokes and resurfaced. The sea, disconcertingly transparent, tasted only vaguely of salt, but the cool water calmed his nerves. Swimming out toward the breakers, he admired the colors of the Jovian planet as it climbed higher above the mountains, its bright green and golden bands becoming richer and better defined as the sun reddened in the west.

»Diagnostics complete. No errors detected.«

"Fantastic!" He flipped onto his back, water lapping around him. He gazed skyward and closed his eyes. "Record the following message into persistent storage, then squirt it real time over the link, slowed by a factor of 30.017."

»Persistent storage online. Ready to record.«

"Doctor Nolen, Céline. Sorry for the silence; unfortunately realtime communication isn't practical." Giddy with excitement, he couldn't help grinning. "You guys are not going to believe this. There's a 30 to 1 time differential in my favor. That means I have roughly three hours to spend in the simulation enjoying the sun and sand while you sit in that dreary lab watching me snooze." Cal laughed. "A speedup of thirty. Think of it! To experience a month of life in a single day. This is so much cooler than we ever imagined."

Autonomy
Reviews

"**THUMBS UP!** Very entertaining. The ideas that he proposes are absolutely fascinating … and the base-60 system, it's **FASCINATING!**"

—*Joey's Culture Corner, TrekWest 5*

"**FAST PACED** … some of the best sci-fi I've ever read!"

—nmatrix9, *Reddit.com*

"**COMPELLING** story … a **SKILLFUL** writer … **I WAS IMMEDIATELY ENTRANCED.**"

—Geoff Lehr, *Blogspot.com*

"**ABSOLUTELY LOVED IT!** There are some books you've got to read over again because you liked [them] so much. This is one of those books.

—NitroNorm, *Amazon.com*

"**SOLID CRAFTSMANSHIP** … engaging … I found myself eager to find out what happens next."

—*SF Signal*

ABOUT THE AUTHOR

JEAN-MICHEL SMITH was born in Palo Alto, California. Pursuing degrees in physics and engineering, he ultimately chose computer science and graduated with a BS from the University of Illinois, College of Engineering. After completing the coursework for his master's degree at Illinois State University, he embarked on a career as a systems engineer and enterprise architect. His work has taken him to many countries across the Americas, Europe and Asia. His essays on free culture and collaborative endeavor appear in the French anthology *LOGS: micro-fondements d'émancipation sociale et artistique*. He is the author of the science fiction novel *Autonomy* and lives with his wife, Christine Todd, in Chicago.

Visit his websites:

jean-michel.eu
autonomyseries.com

Printed in Great Britain
by Amazon